S0-BFA-167

Protecting the Oceans (Crabtree) (Library Binding)   2/12/2021

Titles in set: 4

Total for set: $82.80

In its Sustainable Development Goals for 2030, the United Nations states a healthy Earth requires healthy oceans—and life below water

Effects of Climate Change on the Oceans $20.70

Climate change is one of the most serious threats to the world's oceans causing ecosystem damage, coastal erosion, and ocean acidification. This necessary title explores these effects and the worldwide...

#2175033   N. Hyde   Available:08/31/2020   32 pgs
Grade:345   Dewey:551.6   LEX:IG 1030

Preventing Ocean Pollution $20.70

Most people have heard of the massive collection of debris in the Pacific Ocean or the damaging effects of oil spills on ocean ecosystems. But there are many other ways the world's oceans are being aff...

#2175034   N. Hyde   Available:08/31/2020   32 pgs
Grade:345   Dewey:551   LEX:IG 980

Using Ocean Resources Sustainably $20.70

From food to economic opportunities, the ocean contains a wealth of natural resources that humans around the world depend on. But as we use these resources, we can negatively impact the ocean. Luckily,...

#2175035   N. Hyde   Available:08/31/2020   32 pgs
Grade:345   Dewey:333.91   LEX:IG 1040

Why Are Oceans Important? $20.70

Why are oceans some of Earth's most important resources? This title explores the many ways humans and other living things depend on oceans, from helping to clean the air and regulating the climate to p...

#2175036   N. Hyde   Available:08/31/2020   32 pgs
Grade:345   Dewey:551   LEX:IG 940

Protecting the OCEANS

Using Ocean Resources
SUSTAINABLY

By Natalie Hyde

CRABTREE
PUBLISHING COMPANY
WWW.CRABTREEBOOKS.COM

CRABTREE
PUBLISHING COMPANY
WWW.CRABTREEBOOKS.COM

Author: Natalie Hyde

Editorial director: Kathy Middleton

Editor: Janine Deschenes

Proofreader: Wendy Scavuzzo

Design: Margaret Salter

**Production coordinator
and Prepress technician:**
Margaret Salter

Print coordinator: Katherine Berti

Photo Credits:

b=Bottom, t=Top, tr=Top Right, tl=Top Left

NOAA Fisheries: p12-13 (middle)

Shutterstock: zaferkizilkaya, p5 (t);
Syndromeda, p10 (inset); WoodysPhotos,
p12 (b); pisaphotography, p14 (t); OVKNHR,
p17 (t); Tiew.Zog.Zag, p23 (t); Ink Drop, p27;

All other images from Shutterstock

Library and Archives Canada Cataloguing in Publication

Title: Using ocean resources sustainably / Natalie Hyde.
Names: Hyde, Natalie, 1963- author.
Description: Series statement: Protecting the oceans |
  Includes bibliographical references and index.
Identifiers: Canadiana (print) 20200283790 |
  Canadiana (ebook) 20200283804 |
  ISBN 9780778782032 (hardcover) |
  ISBN 9780778782070 (softcover) |
  ISBN 9781427126078 (HTML)
Subjects: LCSH: Ocean—Juvenile literature. |
  LCSH: Marine resources conservation—Juvenile literature. |
  LCSH: Nature—Effect of human beings on—Juvenile literature.
Classification: LCC GC1016.5 .H93 2020 | DDC j333.91/6417—dc23

Library of Congress Cataloging-in-Publication Data

Names: Hyde, Natalie, author.
Title: Using ocean resources sustainably / Natalie Hyde.
Description: New York : Crabtree Publishing Company, [2021] |
  Series: Protecting the oceans | Includes index.
Identifiers: LCCN 2020029736 (print) | LCCN 2020029737 (ebook) |
  ISBN 9780778782032 (hardcover) |
  ISBN 9780778782070 (paperback) |
  ISBN 9781427126078 (ebook)
Subjects: LCSH: Ocean--Juvenile literature. |
  Marine resources conservation--Juvenile literature. |
  Nature--Effect of human beings on--Juvenile literature.
Classification: LCC GC21.5 .H95 2021 (print) | LCC GC21.5 (ebook) |
  DDC 333.91/64--dc23
LC record available at https://lccn.loc.gov/2020029736
LC ebook record available at https://lccn.loc.gov/2020029737

Crabtree Publishing Company
www.crabtreebooks.com   1-800-387-7650

Printed in the U.S.A./082020/CG20200710

Copyright © **2021 CRABTREE PUBLISHING COMPANY.** All rights reserved. No part of this publication may be reproduced, stored in a retrieval system, or be transmitted in any form or by any means, electronic, mechanical, photocopying, recording, or otherwise, without the prior written permission of Crabtree Publishing Company. In Canada: We acknowledge the financial support of the Government of Canada through the Canada Book Fund for our publishing activities.

Published in Canada
Crabtree Publishing
616 Welland Ave.
St. Catharines, Ontario
L2M 5V6

Published in the United States
Crabtree Publishing
347 Fifth Ave
Suite 1402-145
New York, NY 10016

Published in the United Kingdom
Crabtree Publishing
Maritime House
Basin Road North, Hove
BN41 1WR

Published in Australia
Crabtree Publishing
3 Charles Street
Coburg North
VIC, 3058

CONTENTS

A DIMINISHING RESOURCE

The people of Yumingzui Village in China have always made their living from the Yellow Sea. They fished for flounder, herring, and yellow croaker. But now, with **fish stocks** decreasing and the ocean environment being damaged from fishing, the government took action. It wanted to encourage people to leave fishing and find other jobs. They banned any fishing in the summer. But the local people snuck out to fish by moonlight to pay their bills. Now, with the fishing industry no longer driving the economy, the government has plans to demolish the buildings and build a luxury tourist resort. Villagers say that even if they turn to tourism to create jobs and money, tourists won't come if there is no seafood to eat.

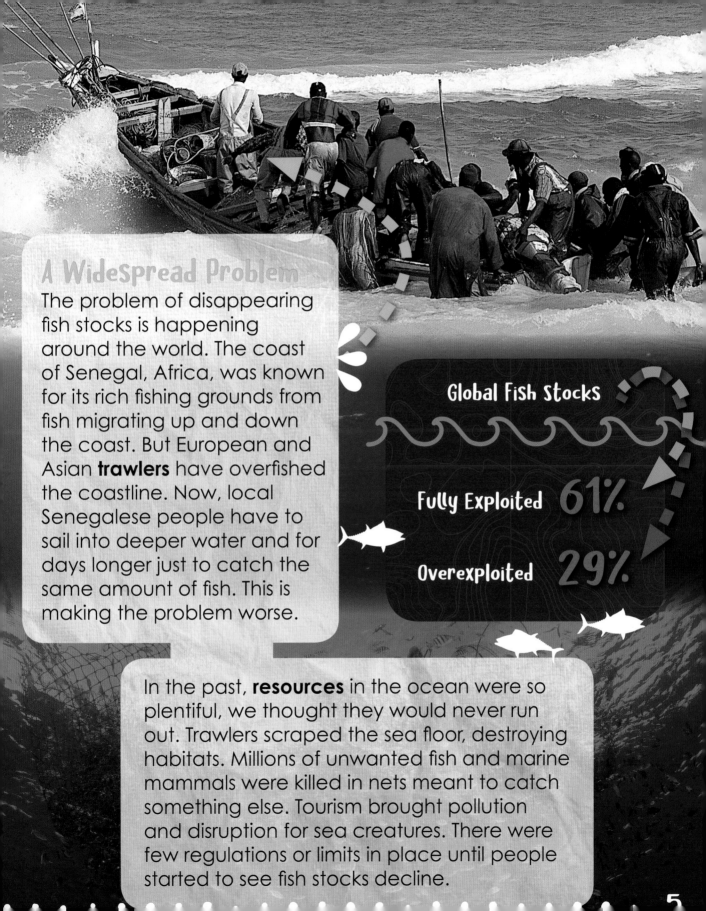

A Widespread Problem

The problem of disappearing fish stocks is happening around the world. The coast of Senegal, Africa, was known for its rich fishing grounds from fish migrating up and down the coast. But European and Asian **trawlers** have overfished the coastline. Now, local Senegalese people have to sail into deeper water and for days longer just to catch the same amount of fish. This is making the problem worse.

Global Fish Stocks

Fully Exploited **61%**

Overexploited **29%**

In the past, **resources** in the ocean were so plentiful, we thought they would never run out. Trawlers scraped the sea floor, destroying habitats. Millions of unwanted fish and marine mammals were killed in nets meant to catch something else. Tourism brought pollution and disruption for sea creatures. There were few regulations or limits in place until people started to see fish stocks decline.

THE NEED FOR SUSTAINABILITY

Our five oceans are connected into one World Ocean. It covers 70 percent of our planet. It is full of life, including marine plants, shellfish, huge whales, sea turtles, and fish of every kind. We have relied on our ocean to provide us with many resources related to food, recreation, mining, and transportation.

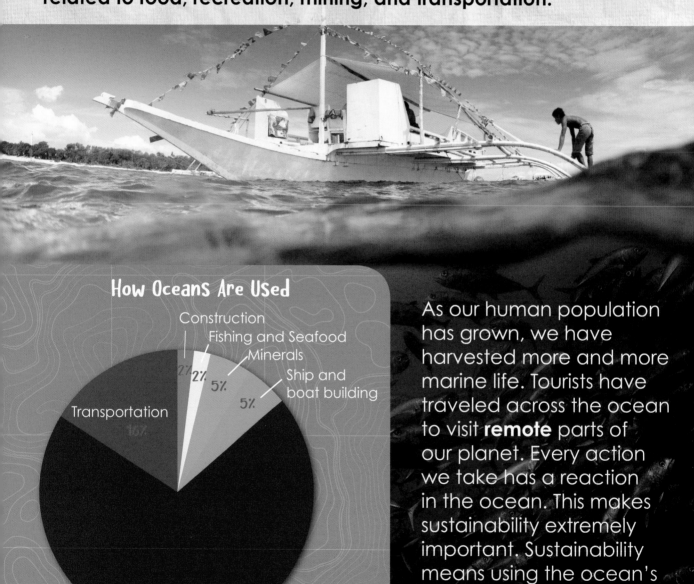

How Oceans Are Used

- Construction 2%
- Fishing and Seafood 2%
- Minerals 5%
- Ship and boat building 5%
- Transportation 16%
- Tourism 70%

As our human population has grown, we have harvested more and more marine life. Tourists have traveled across the ocean to visit **remote** parts of our planet. Every action we take has a reaction in the ocean. This makes sustainability extremely important. Sustainability means using the ocean's resources in a way that keeps them healthy for the future.

OCEAN ACTION

Seychelles

The Seychelles is a group of islands off the east coast of Africa. As part of its commitment to the 14th United Nations Sustainable Development Goal of "Life Below Water," it is expanding marine protected areas. These are designated areas protected by laws. This will help preserve **biodiversity** and allow fish stocks to rebuild.

A Better Future

The United Nations (UN) is a collection of almost 200 countries that work together to solve problems that affect all of us on Earth. In 2015, the UN developed 17 global goals to work toward creating a more sustainable world. Goal 14 is called Life Below Water. It is concerned with finding ways to protect oceans and use their resources sustainably. This is done through such things as finding money to support scientific studies and projects, and creating international laws or treaties. Treaties , or formal agreements, such as the Law of the Sea Convention identify nations' rights and responsibilities in how they use oceans and their resources.

14 LIFE BELOW WATER

TOWARD A BLUE ECONOMY

The health of our oceans and the people using the resources in them is linked to tackling climate change, managing fish stocks, and properly disposing of our waste. Being able to use, harvest, and mine the oceans while keeping them clean and allowing ecosystems to thrive is called the "blue economy."

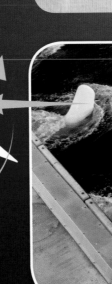

These turbines are turned by **tides** to create electricity.

The need to keep our ocean resources sustainable is not just a responsibility for governments and industry. It is also important for the public to help make decisions and plans. The blue economy includes many ocean activities. **Renewable energy** such as ocean wind farms and tidal stations reduces our use of **fossil fuels** that contribute to **climate change**. People can urge their governments to expand their use of this kind of renewable energy.

OCEAN ACTION

Commonwealth Blue Charter

The Commonwealth is the name given to the countries that were all territories of the **British Empire** in the past. In 2013, all 54 member states signed the Commonwealth Blue Charter. This agreement allows all countries to work together for ocean protection and industry development. It includes projects such as protecting coral reefs, creating marine protected areas, and a sustainable blue economy.

Need for Change

Tourism can benefit developing island countries or coastal communities, but it needs a plan that will not damage or pollute the ocean they depend on. Marine transportation, including moving people and cargo, needs to be safe and non-polluting. The fishing industry is vital to the food supply and economies of many countries around the world. But overfishing is leading to the loss of many fish stocks. The oceans hold much of the carbon dioxide that human activities produce and marine plants produce 70 percent of the oxygen in our atmosphere.

80% of all trade in goods uses ocean transportation.

OVERFISHING PROBLEM

Each year, the amount of fish and other seafood we use for food continues to rise. This has led those who fish for a living to also catch smaller and younger fish to increase their catch. The result is not enough fish grow to be adults that can reproduce. In the last 80 years, fish populations have continued to fall worldwide.

Overharvesting marine plants is having the same effect. Many types of seaweed such as kelp take two to four years to regrow a **bed**. In Maine, harvests of brown algae called rockweed from the 1990s to 2013 exploded from 490 tons (450 metric tons) to 8,488 tons (7,700 metric tons). Rockweed is not only slow growing, but it is a habitat for 150 different fish species. Overfishing and overharvesting can be controlled with new regulations that limit how many can be taken each year.

Local people in Zanzibar gather seaweed every morning when the tide is low. It's sold to China and used in both the food and cosmetic industries.

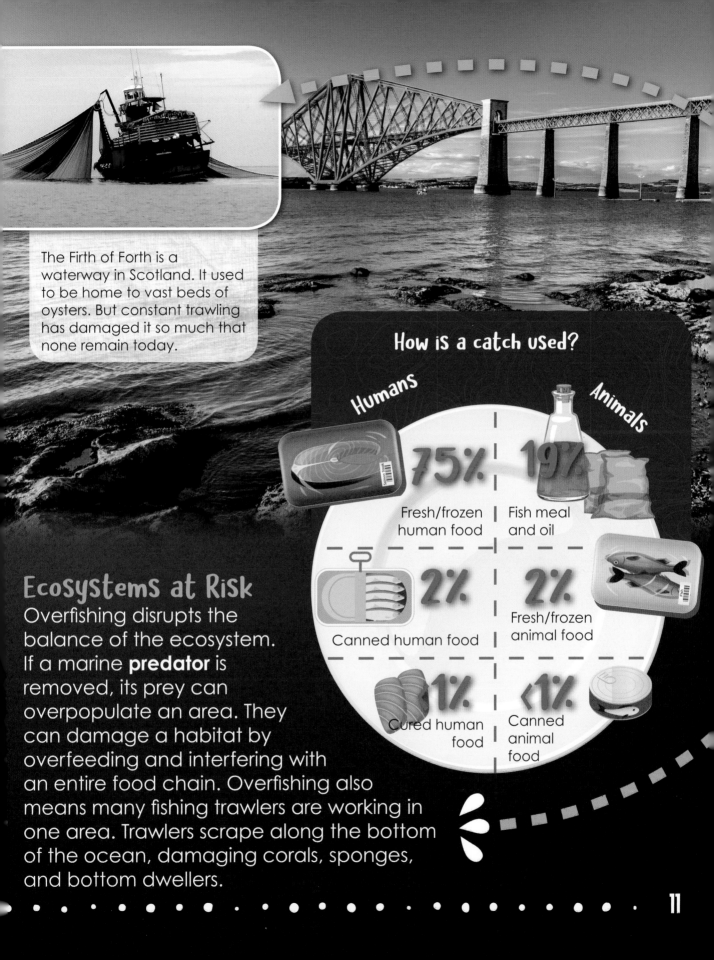

The Firth of Forth is a waterway in Scotland. It used to be home to vast beds of oysters. But constant trawling has damaged it so much that none remain today.

How is a catch used?

Humans

75% Fresh/frozen human food

2% Canned human food

1% Cured human food

Animals

19% Fish meal and oil

2% Fresh/frozen animal food

<1% Canned animal food

Ecosystems at Risk

Overfishing disrupts the balance of the ecosystem. If a marine **predator** is removed, its prey can overpopulate an area. They can damage a habitat by overfeeding and interfering with an entire food chain. Overfishing also means many fishing trawlers are working in one area. Trawlers scrape along the bottom of the ocean, damaging corals, sponges, and bottom dwellers.

UNINTENDED CATCHES

As fishing boats work to catch fish, other creatures get caught in nets, traps, or on fishing lines. Researchers estimate that up to 40 percent of what is caught is unintended. Millions of sea creatures are accidentally caught in this way each year.

More than 50,000 endangered sea turtles may die every year tangled in nets meant for shrimp. Some fishing gear is being adapted by adding updates like trap doors for turtles called TEDs (Turtle Excluder Devices). These are metal grates in trawl nets that allow them to escape and they are 97% effective.

OCEAN ACTION

Spencer Montgomery

Spencer Montgomery calls the precut fish fillets found in supermarkets "faceless fish." This kind of fish was all he knew growing up. But now he is working to bring sustainable seafood to campuses all over the northeastern United States and Canada. He is involved in an organization called Slow Food, which works to preserve food traditions and fight against fast food, which can be unsustainable. Spencer is trying to connect youth to fishermen and the process of preparing fresh food.

Algae blooms can use up all the oxygen in the water, killing fish.

Discards can reach up to
80% or even **90%**
of the total catch.

Dumping Disaster

Unintended catch usually results in the creatures being injured or killed. Researchers estimate that around 24.5 million tons (27 million metric tons) of seafood are scooped up each year. While some is used to make fishmeal for feeding farm animals, the rest is dumped overboard. Dumping dead fish, guts, and carcasses overboard brings **scavengers** that might not normally go to the area. This can affect the normal ecosystem and put stress on the creatures that live there. Dumping unintended catch also puts more nutrients in one concentrated area. This can lead to dangerous algae blooms, which can release toxins into the water.

MAKING SUSTAINABILITY LAW

Laws and regulations are an important way to ensure that ocean resources are used sustainably. These range from protecting ecosystems, limiting fishing, regulating mining and dumping activities, and more. A sustainable future is only possible if everyone follows these laws and regulations.

Companies that choose to mine without approval from the International Seabed Authority (ISA) would be able to use damaging mining practices. Some cruise ships have been caught dumping plastic and food waste into the ocean. One cruise line was fined 20 million dollars for polluting the ocean by illegally dumping oil.

OCEAN ACTION

Kerstin Forsberg

Kerstin Forsberg is working with Planeta Océano, a not-for-profit organization in South America protecting ocean environments. Their goal is to get legal protection for Peru's giant manta rays. Forsberg got tourists and local fishermen to collect data to use in their case. A ban on manta ray fishing was passed in Peru in 2015. Now she wants to create plans for sustainable tourism.

Illegal fishing is a problem in protected areas, such as this marine national park in Thailand. A fishing net covers a coral reef, damaging it and the marine creatures that call it home.

Stopping IUU Fishing

One of the biggest threats to sustainability is Illegal, Unreported and Unregulated (IUU) fishing. It is the biggest reason fish stocks are decreasing. IUU happens within national waters and on the high seas. Researchers estimate that 30 percent of all fishing activity worldwide is illegal. Up to 24 million tons (28 million metric tons) of fish are illegally caught each year. This is a bigger problem near countries with fewer regulations and enforcement, such as off the coast of sub-Saharan Africa. The best way to prevent IUU is to create and enforce stronger international laws and policies. New ways of identifying legal fishing boats and monitoring their locations can also help.

100 million sharks are killed illegally every year.

33% of shark fins for sale in Hong Kong were from threatened species.

MINING THE OCEAN FLOOR

Metals and minerals such as nickel, copper, platinum, and even diamonds have been found on and under the ocean floor. We use them for everything from cell phones to aircraft engines. With mines running out of materials on land, people are looking to mine the ocean floor. So far, deep-sea mining has not begun on big scale. Companies are still researching how profitable this type of mining can be.

Any type of ocean floor mining will damage sea floor structures and marine ecosystems. Silt and sand will be moved and may cover feeding and breeding grounds. Machinery will produce light and noise pollution in a world that is normally dark and quiet. Waste from drilling platforms and ships will pollute the area. These problems will multiply as more and more companies begin to deep-sea mine.

Blue crabs are one of many marine creatures that live on the ocean floor.

Near **hydrothermal vents** (left), copper, gold, and silver have been found. But these areas are also habitats for marine creatures and need to be protected.

An Ocean Mining Code

Mining off a nation's coast will be regulated by that country's government. The concern is who will regulate and enforce rules in the open ocean. The International Seabed Authority (ISA) is a group of 168 countries and was created by the United Nations. The ISA was established to create and control deep-sea mining activities and to share data and scientific studies. Working with the member countries, it has created a "Mining Code." This is a set of rules and regulations that will hopefully achieve sustainable use of marine mineral resources.

One way to help is to recycle items containing in-demand metals and minerals, such as electronics. Recycling reduces the need for mining.

The demand for copper will rise **341%** from 2010 to 2050.

OFFSHORE DRILLING

Metals and minerals are not the only valuable resources on or in the ocean floor. Oil and gas deposits **are also found in deep water. We use them to power our vehicles, furnaces, and power plants.**

Oil and natural gas is accessed using drills from drilling platforms or rigs. But building offshore platforms creates pollution and damage to ecosystems. Once platforms are in place, there is a risk of fires and spills, which can destroy habitats and kill wildlife.

The oil and gas from these structures needs to get to land where it can be processed and used. This is usually done with pipelines. Pipelines can disrupt the migration, feeding, and habitats of sea life in the area. Pipelines are also at risk for damage from storms and strong waves, leading to even more oil spills. To help with this problem, scientists have created robots that can travel through the pipes, cleaning them and checking them for weak spots and cracks. Underwater remotely operated vehicles (ROVs) can also inspect the pipelines from the outside.

More sustainable offshore drilling might involve making a plan that thinks carefully about vulnerable marine areas.

Each year, more than 500,000 birds die due to oil spills.

The biggest oil spill from a drilling rig was the Deepwater Horizon oil spill on April 20, 2010. Size at one point was

15,300 square miles (40,000 sq km)

This is equal to 10 times the size of Rhode Island.

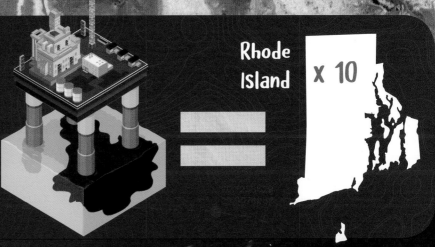

Rhode Island x 10 =

A Safer Industry

Offshore drilling companies have made changes to make the industry safer. They have developed a stronger cement to build the wells that go thousands of feet into the ocean floor. They have also increased the number of remotely operated submarines. These robot submarines work on the ocean floor where it is too dangerous for divers. New systems that use **satellites** and **radar** to track the size and movement of oil spills help crews remove and clean up the oil spill more quickly.

LIGHT AND NOISE

No sunlight can reach deep into the ocean. The creatures there live in a world of darkness. They use other senses to get around, find food, and signal danger. But sounds and light from human-made vehicles create noise and light pollution that can harm them. Using ocean resources sustainably means limiting the effects of noise and light pollution.

Boat engines, submarines, oil drilling, and **sonar** all create sound. Sound travels faster underwater than above. Noise pollution interferes with how well marine creatures hear. The more traffic there is, the more noise pollution is created. Between 1950 and 2000, ship traffic on our oceans doubled.

Some creatures create light with chemicals in their bodies. This is called bioluminescence. They use this light to lure food or scare off attackers. But when underwater vehicles light up their world, they cannot use this valuable tool. Polluting oceans with noise and light affects marine creatures' abilities to grow and reproduce. We cannot continue to have healthy populations of fish and other ocean animals if we disrupt how they live.

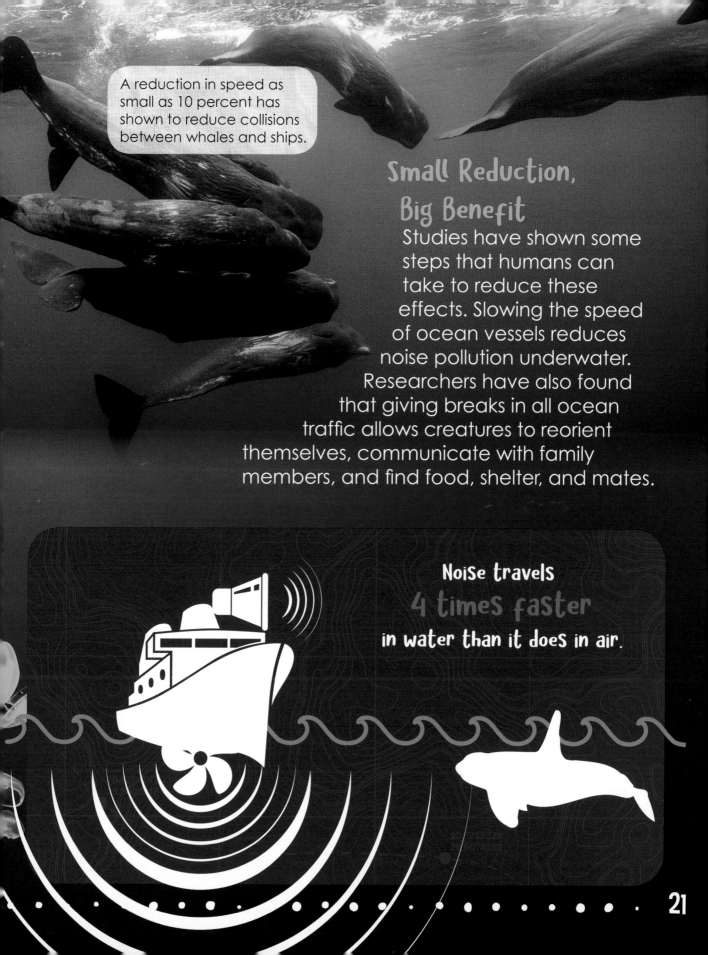

A reduction in speed as small as 10 percent has shown to reduce collisions between whales and ships.

Small Reduction, Big Benefit

Studies have shown some steps that humans can take to reduce these effects. Slowing the speed of ocean vessels reduces noise pollution underwater. Researchers have also found that giving breaks in all ocean traffic allows creatures to reorient themselves, communicate with family members, and find food, shelter, and mates.

Noise travels **4 times faster** in water than it does in air.

NEED FOR ECOTOURISM

Oceans are enjoyed by people around the world. Sports such as diving, surfing, swimming, sailing, and water-skiing are just some of the ways we use the oceans for recreation. Human activities, even for fun, can have a huge impact on ocean creatures and ocean habitats.

Trying to find a balance between visitors enjoying a new country, sport, or environment while also reducing damage to our oceans is a difficult task.

Water craft such as jet skis, speed boats, fishing boats, and seaplanes can create noise, light, and waste pollution. Diving and boating in or over sensitive ecosystems such as coral reefs and **mangroves** can also cause damage. People use oceans to travel and to visit other countries. Tourism can include cruise ships, sailing vacations, houseboats, fishing trips, and diving expeditions. Many people also depend on the tourism industry for their income.

Ecotours often include a guide who can explain the challenges and solutions of the area being visited.

Sustainable Travel

Ecotourism means travelling to natural areas without damaging them. It also means sustaining the well-being of local people who depend on the tourism industry. Today, marine ecotours are being offered in many sensitive areas. Small island nations and coastal countries can benefit from ecotourism. Tourism money can be directed to conservation programs and projects to protect threatened species. Education is also an important part of ecotourism.

A week long cruise dumps around

150,000 GALLONS

(567,812 l) into the ocean each week. That's equal to about

10

backyard swimming pools.

MARINE PROTECTED AREAS

Overfishing, mining, drilling, and tourism all damage and disrupt marine environments. Researchers have found that one of the most effective ways to save ocean ecosystems and the creatures that live in them is to create marine protected areas (MPAs). MPAs shelter ocean life and their habitats. MPAs have limits on human activities such as diving, fishing, mooring boats, and removing marine life.

There are different levels of MPAs. The highest level is a marine reserve. It has legal protection from fishing or development. The most effective marine reserves are large enough to contain habitats for all stages of life of certain species. They will have **spawning** grounds, nursery areas, **home ranges**, and migration routes. MPAs within a country's coastal waters can be created by that country's government. Establishing MPAs in international waters, also called the high seas, is more difficult because no one country has the responsibility of maintaining or enforcing this area.

This marine protected area in Indonesia is home to delicate coral reefs and other habitats.

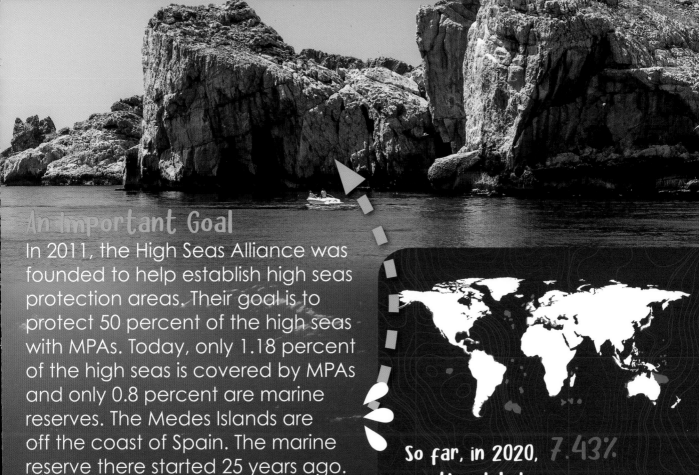

An Important Goal

In 2011, the High Seas Alliance was founded to help establish high seas protection areas. Their goal is to protect 50 percent of the high seas with MPAs. Today, only 1.18 percent of the high seas is covered by MPAs and only 0.8 percent are marine reserves. The Medes Islands are off the coast of Spain. The marine reserve there started 25 years ago. Today, several species of fish living there, such as the dusky grouper, have recovered their populations.

So far, in 2020, 7.43% of the global ocean is marine protected areas

OCEAN ACTION

Marine Megafauna Foundation

The Marine Megafauna Foundation (MMF) focuses on research and conservation for threatened marine **megafauna**. These include sharks, rays, marine mammals, and turtles. The organization works to improve the management of existing MPAs. They help develop effective, long-term conservation strategies to protect and restore key habitats.

GET INVOLVED

As the world's population grows, we will need the ocean to produce more food, jobs, and energy. It is essential that we work together now to use the ocean's resources more sustainably so that future generations can meet their needs.

Sustainable use means leaving enough adult fish to reproduce for the following year's catch. It means mining without destroying seafloor habitats. It also means changing and limiting ocean travel to not increase pollution or disturb communication for marine creatures. We also need to limit or reverse climate change to stop the warming and **acidification** of our oceans.

Acidification is caused by too much carbon dioxide in the ocean. Among other effects, it causes marine creatures, such as coral, to have difficulty growing hard shells and skeletons.

USe Your Voice

We can all do our part to help. Everyone has a voice and it is the job of government to listen to citizens. We can write to our representatives and ask them to support the UN's Sustainable Development Goals and other proposals that will protect our oceans. Raising awareness for these goals can include joining activities or giving support for international policies and agreements such as the Paris Agreement on reducing climate change. We can keep our governments accountable by asking questions and putting pressure on them to do their part through protests and petitions. Norway is one country that is beginning a plan that combines green technology and the fight against illegal fishing to work toward sustainable ocean use. We can also ask our schools, businesses, and restaurants to buy products from sustainable sources, and lower their **carbon footprint** and plastic use.

CREATING A BETTER FUTURE

From making informed food choices to limiting waste, practicing ecotourism, and educating ourselves about climate change, we can also make decisions in our own lives that affect the oceans and keep them sustainable.

When we shop or eat out, we can make choices about what seafood to eat. Experts say locally caught fish is best. That means it has not had to travel far to get to your plate. Transporting food products uses fossil fuels, which contribute to climate change. We can also help fight climate change and its effects on oceans by supporting renewable energy projects. You can write to your government representatives and ask them to push for renewable energy such as tidal or wind farms. You can also educate yourself and others about climate change. All of these actions help create a better future.

When shopping, look for labels that identify seafood that has been caught following sustainability guidelines.

OCEAN ACTION

Nina Jensen

Nina Jensen is the CEO of REV Ocean. REV Ocean is the world's largest research and expedition vessel. It houses 60 researchers at a time. Its goal is to find sustainable and environmentally responsible solutions to the problems of the oceans. Jensen believes that, by working together, we can have the most success at developing new ways to protect our oceans.

REV Ocean

You Can Help

Keeping our oceans healthy for marine life to thrive can start at home. We can choose to use alternatives to plastic, get rid of our waste properly, and take eco-friendly vacations. We can also join beach cleanups or support organizations that are committed to preserving and protecting ocean life. It is also important to keep up to date on what is happening in our oceans. Use terms such as "ocean conservation," "marine protected areas," or "sustainable oceans" in search engines. Try to get a balanced view on what is happening. Knowing what is happening in our oceans will help you make good decisions about

When researching, check for government websites as well as not-for-profit organizations to double check facts.

GLOSSARY

acidification Increasing amount of acid

bed An area of ground or ocean floor where plants grow

biodversity Number of different plants and animals that live in an area

British Empire All the countries and territories that were once controlled by the United Kingdom

carbon footprint Amount of carbon dioxide released into air from fossil fuel use by one person or business

climate change The change in the world's weather patterns due to higher levels of the gas carbon dioxide in Earth's atmosphere

deposits Layers or bodies of accumulated matter

ecosystem A community of organisms and their environment

exploited Caught only up to a number that allows the fish stock to reproduce enough to replace them

fish stocks Population of a type of fish

fossil fuels Fuel formed in the ground long ago from plant or animal remains

home ranges Areas in which certain animals normally travel

hydrothermal vents Openings in the ocean floor that give off hot, mineral-rich water

mangroves Trees or shrubs growing in coastal swamps

megafauna Large animals

predator An animal that hunts and eats other animals for food

radar A system that uses radio waves to find objects and movements

remote Isolated away from other people and places

renewable energy Energy created from a source that doesn't run out

resources Materials that can be used

satellites Human-made objects that orbit Earth and collect data

scavengers Animals that feed on other dead animals

sonar A system that uses sound pulses to locate things

spawning The process of releasing eggs and sperm to reproduce

tides The natural rising and falling of the ocean, which happens twice a day

trawlers Fishing boats that drag a fishing net behind them

unintended Not planned

LEARNING MORE

Books

Berger Kaye, Cathryn and Phillipe; Cousteau. *Make a Splash!: A Kid's Guide to Protecting Our Oceans, Lakes, Rivers, & Wetlands.* Free Spirit Publishing, 2012.

Kurlansky, Mark. *World Without Fish*. Workman Publishing, 2014.

Shea, Therese. *Overfishing*. Gareth Stevens Publishing, 2014.

Websites

http://www.fao.org/fao-stories/article/en/c/1258376/
The Food and Agriculture Organization of the United Nations has ideas on how kids can help save our oceans.

https://fishandkids.msc.org/en/play/all-about-the-oceans
Learn why the sea needs our help with the Marine Stewardship Council.

https://marineprotectedareas.noaa.gov/dataanalysis/ mpainventory/mpaviewer/
https://www.dfo-mpo.gc.ca/oceans/maps-cartes/ conservation-eng.html
Explore these maps to find Marine Protected Areas in the United States and Canada.

INDEX

ABOUT THE AUTHOR

Natalie Hyde has written more than 90 fiction and non-fiction books for young readers. When she gets time to relax, one of her favorite places to be is beside the ocean on a warm, sandy beach.